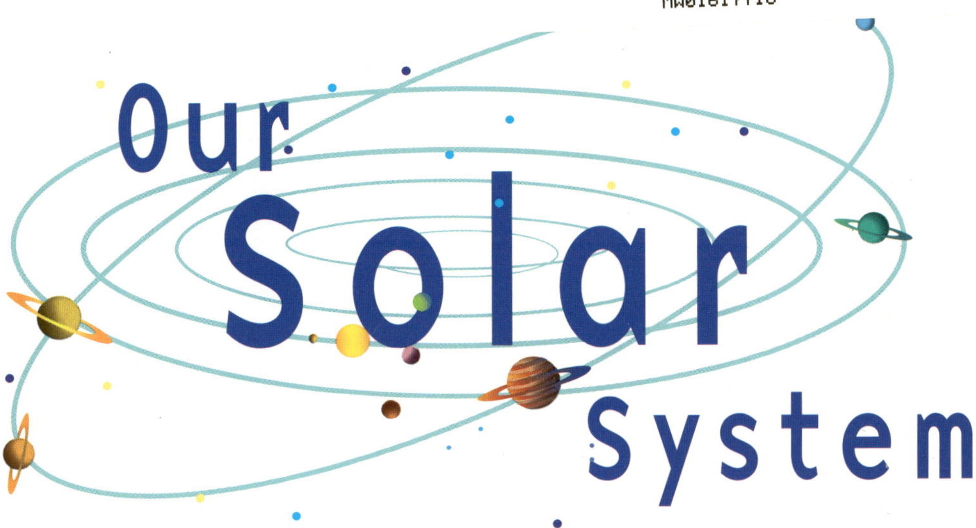

BY RACHEL KRANZ

Table of Contents

What Is Our Solar System?

Chances are you know your neighborhood very well. You probably know a lot about the country you live in, too. You may even know some interesting facts about Earth. But how much do you know about the **solar system**, of which Earth is a member? You are about to take a tour of this vast and remarkable place.

The word "solar" means "about the sun." It comes from the Latin word *sol*, which means "sun." The solar system is the **Sun** and everything that moves around it. It includes:

- the Sun
- the nine **planets** that orbit, or move around, the sun
- the moons that orbit some of the planets
- other objects, including big chunks of rock called **asteroids**, and balls of ice, rock, and dust called **comets**.

It's a FACT!

Our Sun is big and bright—but many other stars are much, much bigger and brighter. Our Sun is considered a medium-size star.

Our Sun, which is the center of the solar system, is really a **star**. It is just like the stars you see in the sky at night, except that it is much closer.

All of the light in the solar system comes from the Sun. It is the only object in our solar system that can produce light by itself. Even moonlight comes from the Sun.

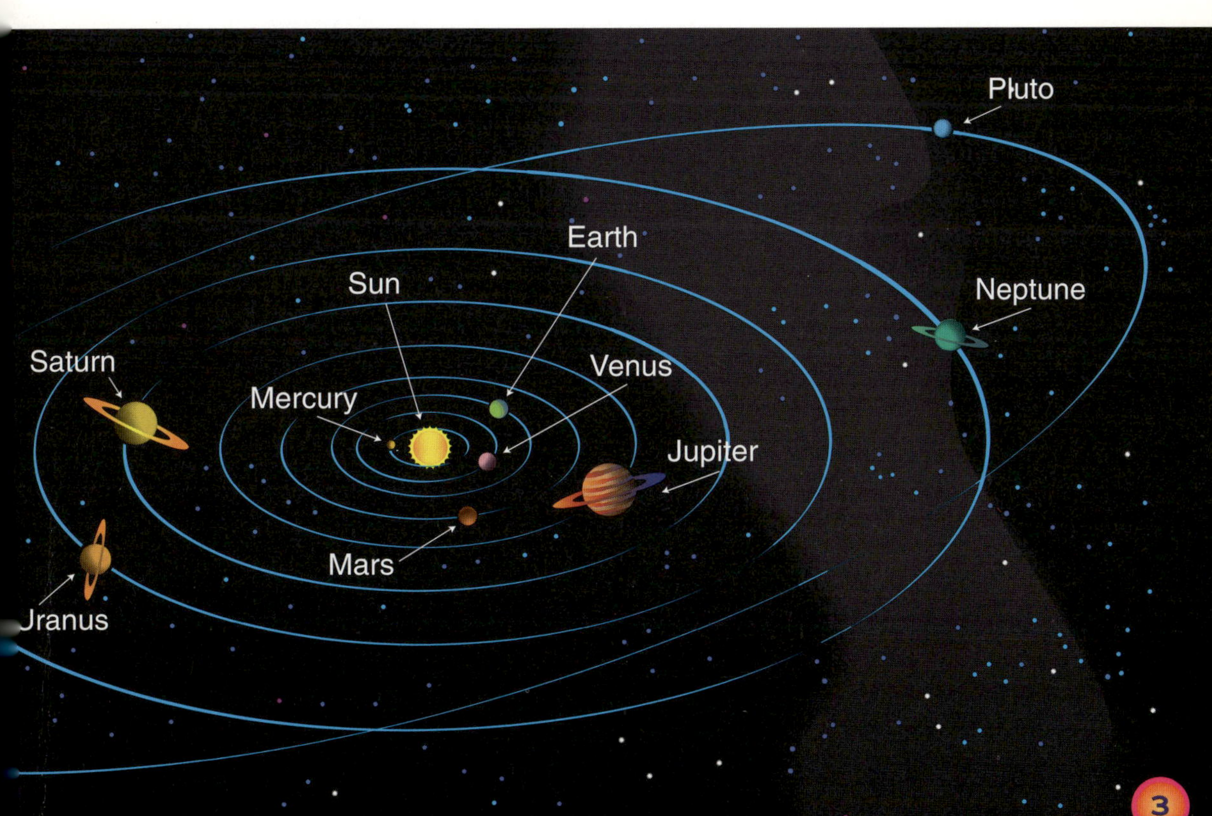

the Sun

Most of the heat in our solar system comes from the Sun, too. The surface temperature of the Sun is about 10,000 degrees Fahrenheit (5,538 degrees Celsius). The inner core, or center, of the Sun is even hotter—27 million degrees Fahrenheit (14,999,982 degrees Celsius)!

Luckily for us, the Sun is very far away—93 million miles, in fact. If it were much closer, Earth would be too hot to support life. If the Sun were any farther away, Earth would be too cold. The Sun is just the right distance away from Earth to support the forms of life that have developed here.

What makes our Sun so hot and bright? Like other stars, the Sun is a huge ball of spinning gases. These gases are **helium** and **hydrogen**. In the core, these gases undergo reactions that produce huge amounts of energy.

Every second, the Sun changes 600 million tons of hydrogen to helium, creating energy. Some of that energy reaches Earth in the form of sunlight. It takes eight minutes for this energy to reach us. Some of the energy reaches other planets. Some disappears into space.

It's a FACT!

The Sun is so hot that if just one spark from its core were to land on Earth, it could set fire to everything within sixty miles of it! The Sun is so enormous that one million Earths could fit inside it.

Nothing on Earth could grow without the Sun's energy.

The nine planets in the solar system all orbit the Sun. They are held in their orbits by the Sun's **gravity**, or force of attraction for them. Each planet's orbit takes a different amount of time. For example, Earth's orbit takes 365 ¼ days. That's why we say that a year on Earth takes 365 days.

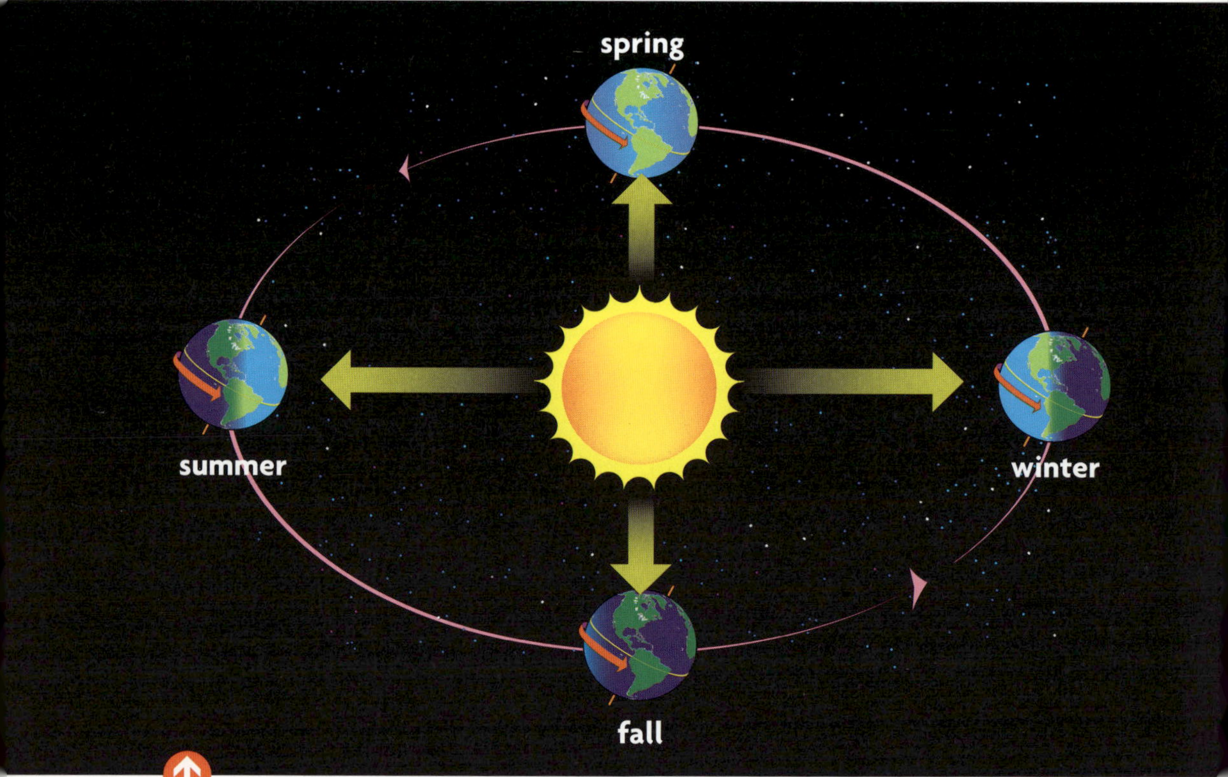

This diagram shows Earth in orbit around the Sun over the course of a year. This movement causes the seasons in many parts of the world. The seasons in the Northern Hemisphere are shown here.

While the planets are orbiting the Sun, they are also rotating, or spinning around an axis like a toy top. One complete **rotation** of Earth takes twenty-four hours, or one day and one night. When our part of Earth is facing the Sun, it is daytime for us. When our part of Earth is turned away from the Sun, it is nighttime. Some planets spin faster than Earth, so their days are shorter. Some spin slower, so their days are longer.

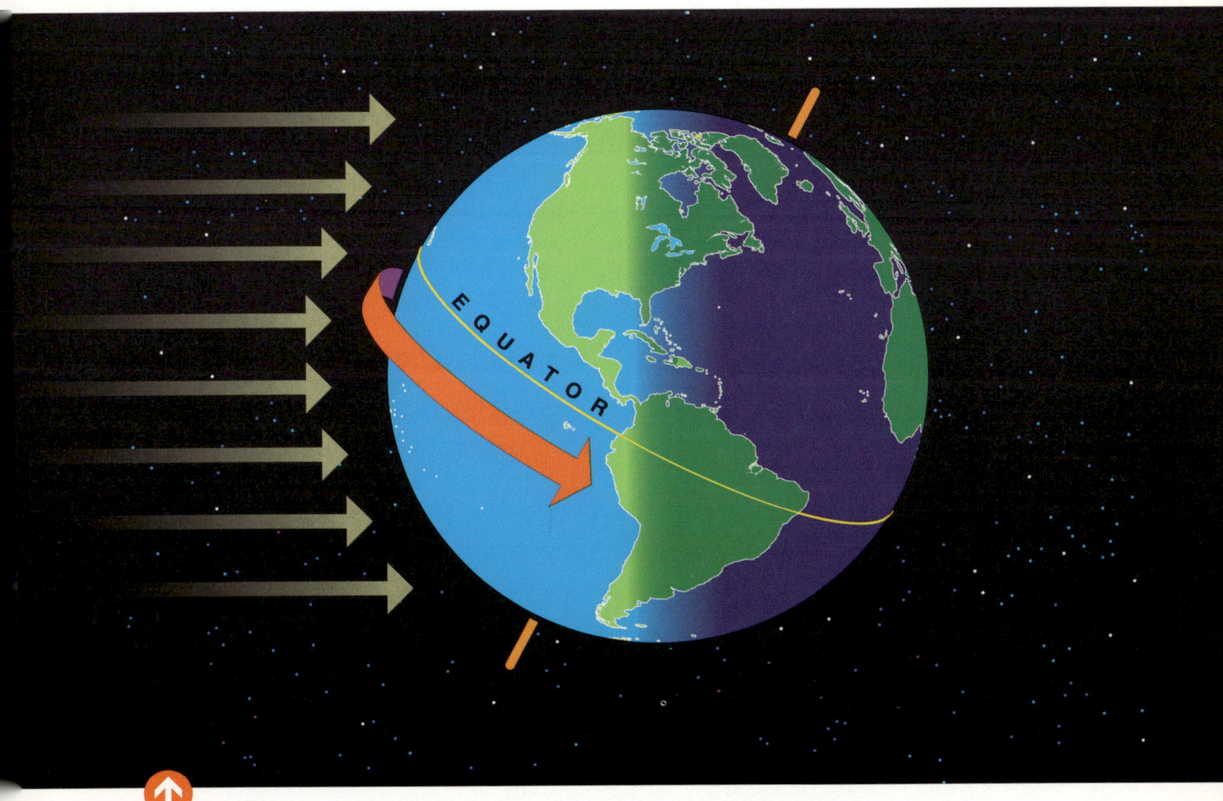

It takes Earth twenty-four hours to rotate once on its axis. What part of Earth is in daytime in this diagram?

CHAPTER 2

The Four Inner Planets: Mercury, Venus, Earth, and Mars

Mercury

Mercury, the planet closest to the Sun, is much smaller than Earth. Earth's **diameter** (the distance through Earth from the North Pole to the South Pole) is 7,900 miles. Mercury's diameter is only 3,000 miles.

Mercury orbits the Sun very quickly. A year on Mercury takes only eighty-eight Earth days. But the planet rotates very slowly. A day on Mercury is fifty-nine Earth days! During the day, the temperature reaches a scorching 806 degrees Fahrenheit (430 degrees Celsius). At night, the temperature drops to a freezing -274 degrees Fahrenheit (-170 degrees Celsius).

This image of Mercury was put together from photos taken by the *Mariner 10* probe.

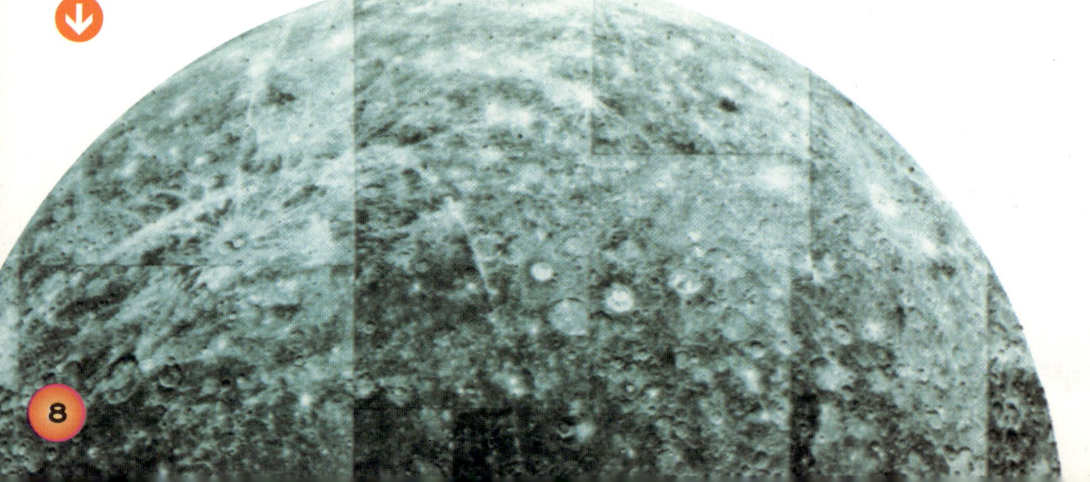

Venus

Venus is the second-closest planet to the Sun. Sometimes you can see Venus in the sky, especially in the early morning or early evening. For this reason, Venus is sometimes called the "morning star" or the "evening star," even though it is not a star at all.

The bright, starry appearance of Venus is caused by the thick layer of clouds around it. These clouds reflect light from the Sun.

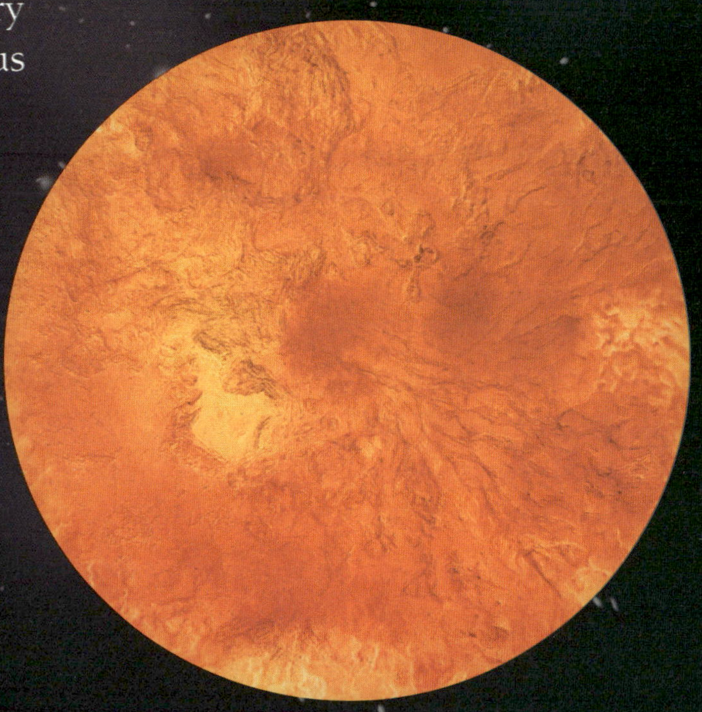

Venus is about the same size as Earth, but the two planets are very different. Venus is very hot, and the lovely clouds that surround it are made of sulfuric acid. Sulfuric acid is a very strong acid that can burn right through metal!

Venus is one of only two planets that rotate in a backward direction. Earth rotates from west to east, but Venus rotates from east to west.

It's a FACT!

In 1982, the former Soviet Union sent a spaceship to explore Venus. There were no people on the ship, only scientific equipment. Radio transmitters beamed back to Earth the first color photos of the surface of Venus. Just two hours later, heat destroyed the transmitters and all the other equipment!

Venus is covered with craters, volcanoes, mountains, and lava plains.

Earth

The third planet from the Sun is our own planet, Earth. It is the only planet in the solar system where life as we know it is found. Life on Earth requires water, oxygen, and temperatures that are not too hot or too cold. Earth is the only planet that has all the right conditions for humans, animals, and plants to survive.

It takes the Moon twenty-nine days to orbit Earth. This amount of time is known as one month. Can you see how the word "month" comes from the word "moon"?

Just as Earth orbits the Sun, a smaller body orbits Earth. This is our Moon. The word "moon" means a body that orbits a planet. All the other planets except Mercury and Venus have moons, too. In fact, most planets have more than one moon.

HOW DID THE INNER PLANETS GET THEIR NAMES?

Many planets in our solar system were named after gods in Greek and Roman mythology.

- Mercury was named for the Roman messenger of the gods. Mercury had winged sandals so he could move quickly. The planet Mercury also moves quickly!
- The shining, beautiful planet Venus was named for the Roman goddess of love and beauty.
- Earth is the only planet whose English name does not come from mythology.
- The red, angry-looking planet Mars was named for the Roman god of war.

Mars

The fourth planet from the Sun is Mars. Mars is called "the red planet" because of its red cliffs and orange sky. Temperatures on Mars can reach 70 degrees Fahrenheit (21 degrees Celsius) during the day. But they drop to -207 degrees Fahrenheit (-133 degrees Celsius) at night.

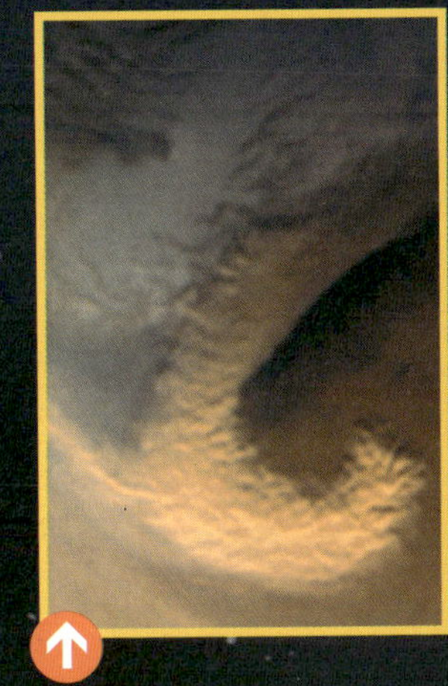

This photo shows a dust storm on Mars.

There are huge windstorms and fierce tornadoes known as "dust devils" on Mars. A dust devil can make a column of dust that is five miles high.

CHAPTER 3

The Five Outer Planets: Jupiter, Saturn, Uranus, Neptune, and Pluto

WHAT ARE THE PLANETS MADE OF?

- The inner four planets—Mercury, Venus, Earth, and Mars—are made of rock. They are called the "terrestrial [Earthlike] planets."
- Four of the outer planets—Jupiter, Saturn, Uranus, and Neptune—are made of gas. They are known as the "gas giants."
- The fifth outer planet, Pluto, is made of ice.

Once you get out to the five outer planets, you are more than 400 million miles away from the Sun. There's not much heat or light that far out in the solar system. But the Sun's gravity is still powerful enough to keep the five planets in their orbits.

This is Jupiter and four of its moons.

Jupiter

Jupiter is the solar system's biggest planet. More than 1,300 Earths could fit inside it.

Because Jupiter is so much bigger than Earth, its gravity is far greater. If you stood on Jupiter, you would weigh two and a half times as much as you weigh on Earth.

Jupiter spins very fast. It rotates once every ten hours compared to Earth, which takes twenty-four hours. Seen through a powerful telescope, Jupiter has gray, brown, blue, and orange stripes swirling around it. These stripes are made of gas. They swirl because the planet is spinning fast. Jupiter has twenty-eight moons. It also has a feature called the Great Red Spot. Scientists think that a large storm, bigger than Earth, causes this spot.

Saturn

The next planet, Saturn, is similar to Jupiter. It is the second largest planet in the solar system. It spins almost as fast as Jupiter.

Pieces of rock and ice spin around Saturn. Some of these pieces are as small as grains of sand, and some are bigger than a house. From a distance, these pieces of rock and ice look like solid rings circling the planet.

For years, Saturn was thought to be the only planet with rings. Then scientists discovered that Jupiter, Uranus, and Neptune have rings, too. However, Saturn has more rings than any other planet. It also has at least twenty-four moons.

Uranus

Almost 2 billion miles away from the Sun is Uranus, the third-largest planet. Uranus looks blue-green because of all the **methane** gas around it.

Like Venus, Uranus rotates in a backward direction, from east to west. It also rotates on such a tilted axis that it looks as if it's lying on its side. Uranus is so far from the Sun that it takes eighty-four Earth years for the planet to make one complete orbit!

HOW DID THE OUTER PLANETS GET THEIR NAMES?

- Jupiter, the biggest planet, was named for the Roman king of the gods.
- Saturn was named for Jupiter's father.
- Uranus was the Roman name for "Father Sky."
- Neptune was named for the Roman sea god.
- Pluto was named for the Greek god of the underworld.

Neptune

Because Neptune is so far away, it has taken scientists a long time to learn about this icy, blue planet.

In 1989, the unmanned spacecraft *Voyager 2* took the first close-up photos of Neptune. That's when we found out that Neptune has rings. We also learned that Neptune has the stormiest weather in the solar system.

Neptune rotates more quickly than Earth. It takes Neptune about eighteen hours to complete a full rotation. But Neptune takes almost one hundred and sixty-five Earth years to complete one full orbit around the Sun.

It's a FACT!

A storm on Neptune can be as large as our entire planet Earth, with freezing winds that blow ten times faster than a hurricane.

Pluto

Pluto is the smallest planet in our solar system. In fact, it's even smaller than Earth's moon!

Although Pluto is usually last on the list of planets, it's not always the farthest from the Sun. Its orbit intersects, or crosses, Neptune's orbit, so that Pluto is sometimes closer to the Sun than Neptune and sometimes farther away.

If you stood on Pluto and looked for our Sun, you would see just a very tiny star. Because Pluto is so far from the Sun, it is the coldest planet. Its average surface temperature is -369 degrees Fahrenheit (-223 degrees Celsius).

Pluto (right) has an unusual moon named Charon that is almost half its size. Charon and Pluto actually revolve around each other. They are known as a double-planet system.

Pluto is a very unusual planet. While the other planets are made of either rock or gas, Pluto is made of a mixture of rocky materials, ice, and other frozen gases.

Some scientists think that Pluto should not be called a planet. Because it's so small, they think that Pluto should be considered part of the Kuiper (KYE-per) Belt, a ring of more than 72,000 small objects that orbit the Sun beyond Neptune.

REMEMBERING THE PLANETS

To remember the names of the planets in order, just say:
Mark's Very Extravagant Mother Just Sent Us Ninety Parakeets.
The first letter of each word is the same as the first letter of each planet, starting with the planet closest to the Sun and moving outward.

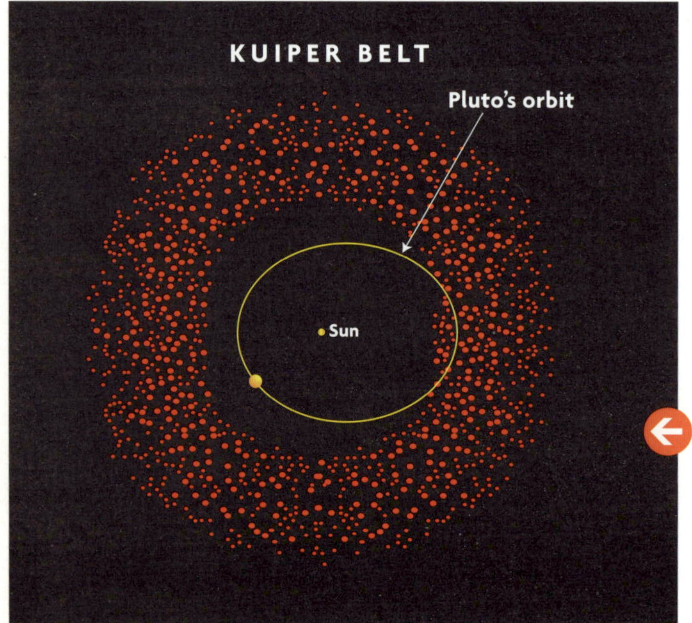

The Kuiper Belt includes Pluto and more than 72,000 other objects, all revolving around the Sun.

CHAPTER 4

Comets, Asteroids, and Other Flying Things

Have you ever seen a science-fiction movie about a gigantic rock from outer space hurtling toward Earth with disastrous consequences? If so, don't worry. The chances of this happening in real life are extremely slim. Still, there are many fascinating objects in our solar system. The Sun's powerful gravity keeps these objects in orbit, just like the planets and their moons.

Meteors are usually easy to see streaking across the sky.

Comets are moving balls of dust, ice, and rock. As a comet gets closer to the Sun, the Sun's heat melts the ice and turns it into gas. This gas forms a long tail that carries pieces of rock and dust.

Like the planets and their moons, a comet has no light of its own. Its light is just sunlight reflecting off the rock and dust. To us, a comet looks like a big, blurry star with a shining tail.

Like planets, comets orbit the sun. That means some of them can be identified and their appearance predicted.

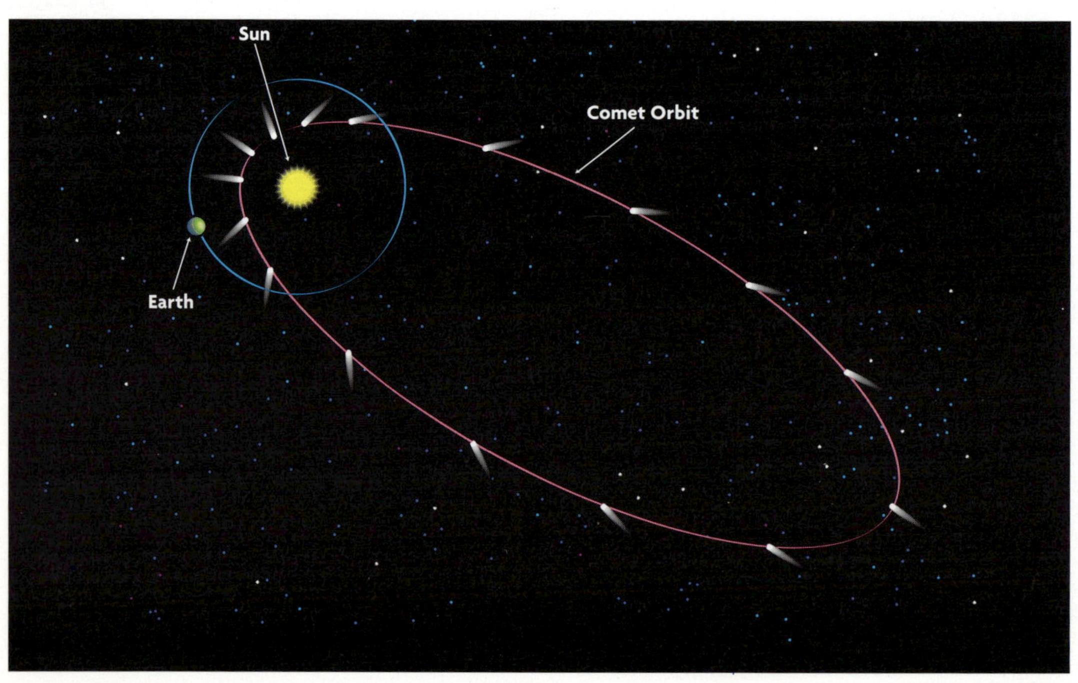

It's a FACT!

A comet's tail can be millions of miles long. Most comets cannot be seen without a telescope. When you do see a comet in the sky, it does not appear to be moving.

This comet was photographed in 1975.

MARK TWAIN AND HALLEY'S COMET

Mark Twain, the famous American author who wrote *The Adventures of Tom Sawyer*, was born on November 30, 1835, in Florida, Missouri. That night, Halley's Comet was clearly visible in the sky. Twain later predicted that he would die the next time Halley's Comet showed up. He wrote: "I came in with Halley's Comet...and I expect to go out with it." And indeed, when Mark Twain died on April 21, 1910, Halley's Comet could clearly be seen overhead.

Long ago, people were frightened by comets. They thought comets were evil signs. If a comet appeared, fortune-tellers might predict that a war would start or a ruler would die.

In 1705, a scientist named Edmond G. Halley discovered that comets appeared regularly in our solar system. He noticed that one particular comet showed up about every seventy-six years. In fact, people had told stories about seeing this comet ever since 240 B.C.

The comet was named Halley's Comet, in honor of the scientist. The last time Halley's Comet appeared was in 1986. Its next visit is expected in 2061.

This photo of Halley's Comet was taken in 1986.

Other chunks of rock and metal that appear in the solar system are called **asteroids**. Thousands of asteroids orbit the Sun in a belt, or ring, between the planets Mars and Jupiter. Although they orbit the Sun, they are too small to be called planets. Scientists believe that asteroids are made of material that remained after the solar system was first formed. However, no one knows for sure where they come from.

The Latin word for "star" is *aster*. The word ending *-oid* means "resembling." So the word "asteroid" means a body that resembles a star. In fact, asteroids don't really resemble stars at all. But to the early scientists who named them, every object in the heavens seemed to be a star.

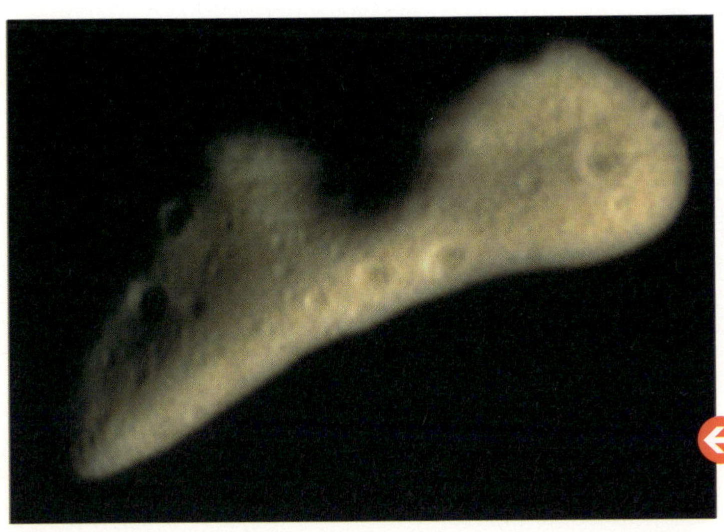

Most asteroids orbit the Sun between Mars and Jupiter. The asteroid shown here is Eros. It was discovered in 1898. It is about 19 miles (32 kilometers) in diameter.

HOW DO ASTEROIDS GET THEIR NAMES?

Usually, the person who discovers a particular asteroid gets to suggest a name for it. Then a decision is made by the International Astronomical Union's Small Bodies Names Committee. The committee tries to pick a name that is easy to pronounce and spell. If a lot of people send e-mails to the committee suggesting one particular name, the committee will consider the name.

Scientists are busily trying to name the asteroids. To do this, they are using the names of science writers (Isaac Asimov, Carl Sagan), composers (Mozart, Bach), and rock stars (Jerry Garcia, Frank Zappa).

This asteroid, named 243 Ida, was the 243rd asteroid discovered.

An asteroid that comes near Earth is called a **meteoroid** (MEE-tee-uh-roid). When a meteoroid shoots through Earth's atmosphere, it becomes so hot that it glows red and burns up. It makes a beautiful streak of light in the dark sky that lasts only a few seconds. A burning meteoroid is called a **meteor**.

Have you ever seen a shooting star? What people call a shooting star is really a meteor.

Most meteors burn up in the atmosphere. A few, however, fall to Earth. A meteor that strikes Earth's surface is called a **meteorite**. When a large meteorite hits Earth, it makes a huge hole, or crater, in Earth's surface.

The Meteor Crater near Winslow, Arizona, was formed when a meteorite fell to Earth about 50,000 years ago.

Sometimes meteors come in groups. A group of meteors is called a meteor shower, because many small bits of rock and metal fall toward Earth at once. The pieces burn brightly as they shoot through the air.

Here is a meteor shower at Arizona's Meteor Crater.

Unanswered Questions

Thanks to scientists, we now know a great deal about our solar system. But there is still much to be learned. There are also some issues that scientists have different opinions about.

- Should Pluto be considered a planet?
- Where did asteroids come from?
- Exactly what causes gravity?

These are just some of the questions that scientists are working to resolve. Perhaps you will be the one to find some of the answers!

Go To

To learn more about the solar system, visit these websites for up-to-date information.

- **The Nine Planets:** seds.lpl.arizona.edu/nineplanets/nineplanets/
- **Spacelink:** spacelink.nasa.gov

Radio telescopes pointed at a starry sky record data for astronomers to study.

Glossary

asteroid	chunk of rock that *orbits* the *Sun*
comet	ball of ice, rock, and dust that *orbits* the *Sun*
diameter	distance across a sphere through its center
gravity	force produced by Earth, the *Moon*, the *Sun*, and other *planets*, *moons*, and *stars*; force of attraction between two objects
helium	type of gas, found at the core of a *star* and other places
hydrogen	type of gas, found on the surface of a *star* and other places; used by a *star* to produce energy
methane	type of gas found around Uranus and other places
meteor	a *meteoroid* that burns as it moves through Earth's atmosphere
meteorite	a *meteor* that falls to Earth
meteoroid	an *asteroid* that comes near Earth
moons	bodies that *orbit* a *planet*
orbit	to move in a fixed path; the path one object follows around another object
planets	large bodies in space that *orbit* the *Sun* and may have one or more *moons*
rotation	one full spin
solar system	the *Sun* and all the bodies that *orbit* it, including the *planets*, their *moons*, *asteroids*, and *comets*
star	gigantic, burning ball of gas that produces light and heat
sun	*star* at the center of our *solar system*

Index